THIS BOOK IS THE PROPERTY OF:

STATE _____	Book No. _____
PROVINCE _____	Enter information
COUNTY _____	in spaces
PARISH _____	to the left as
SCHOOL DISTRICT _____	instructed
OTHER _____	

ISSUED TO	Year Used	CONDITION	
		ISSUED	RETURNED

PUPILS to whom this textbook is issued must not write on any page or mark any part of it in any way, consumable textbooks excepted.

1. Teachers should see that the pupil's name is clearly written in ink in the spaces above in every book issued.
2. The following terms should be used in recording the condition of the book: New; Good; Fair; Poor; Bad.

Ben is a big help.

Ben helps Gram.

Ben sets up the cups.

Zeb gets one. It is a mess.

Ben's mom lets him help.

Ben sets up one net.

Ben's pals get set.

Ben's pals tell him.

"Ben is the best!"

"Ben is ten!"

The End

Understanding the Story

Questions are to be read aloud by a teacher or parent.

1. What is the title?

2. What is Ben helping to do?

3. Do you think Ben knows about his birthday party? Why do you think that?

Answers: 1. Ben Is Ten! 2. get ready for his birthday party 3. no, because his friends surprise him

Saxon Publishers, Inc.
Editorial: Barbara Place, Julie Webster, Grey Allman, Elisha Mayer
Production: Angela Johnson, Carrie Brown, Cristi Henderson

Brown Publishing Network, Inc.
Editorial: Marie Brown, Gale Clifford, Maryann Dobeck
Art/Design: Trelawney Goodell, Camille Venti, Joan Asikainen
Production: Joseph Hinckley

© Saxon Publishers, Inc., and Lorna Simmons

All rights reserved. No part of the material protected by this copyright may be reproduced or utilized in any form or by any means, in whole or in part, without permission in writing from the copyright owner. Requests for permission should be mailed to: Copyright Permissions, Harcourt Achieve Inc., P.O. Box 27010, Austin, Texas 78755.

Published by Harcourt Achieve Inc.

Saxon is a trademark of Harcourt Achieve Inc.

Printed in the United States of America
ISBN: 1-56577-953-3

4 5 6 7 8   446   12 11 10 09 08 07 06

Phonetic Concepts Practiced

ĕ (ten, End)

Nondecodable Sight Words Introduced

one

ISBN 1-56577-953-3

Grade K, Decodable Reader 7
First used in Lesson 74